HOW TO MAKE 7 FIGURE INCOME FROM BULK SMS

CONTENTS

ABOUT THE AUTHOR

CHAPTER ONE:
THE POWER OF THE MOBILE

CHAPTER TWO:
THE CONCEPT OF MOBILE MARKETING

CHAPTER THREE:
BULK SMS EXPLAINED - THE MONEY MAKER

CHAPTER FOUR:
BULK SMS RESELLER PACKAGE

CHAPTER FIVE:
ADVICE

REFERENCE

ABOUT THE AUTHOR

Stephen Akintayo, an inspirational speaker and Serial Entrepreneur is currently the Chief Executive Officer of Stephen Akintayo Consulting International and Gtext Media and Investment Limited, a leading firm in Nigeria whose services span from digital marketing, website design, bulk sms, online advertising, Media, e-commerce, real estate, Consulting and a host of other services. Born In Gonge Area of Maiduguri, Borno State. North Eastern part of Nigeria in a very poverties environment and with a civil Servant as a Mother who raise him and his four other siblings with her mega salary since his father's Contract Business had crumbled. Some of his passion for philanthropy was birthed by his humble Beginning. In his word; "My Surname was poverty. Hunger was my biggest challenge. I had to scavenge all through Primary school to eat lunch as I don't go to school with lunch packs. We were too poor to afford that. Things got better in my secondary school days, though my mum will still go to her colleague to borrow money to send me to school each term. It was humiliating seeing their disdain faces looking at my mum like a foolish woman who keeps begging. It hurts dearly. I hate Poverty and I pray to help more families come out of it". Stephen Akintayo story is indeed a grass to grace one. His singular regret in life is that his hard working mother died few year back due to ovarian cancer and never lived to see some of the good works God is using him to do today.

Stephen, Also Founded GileadBalm Group Services which has assisted a number of businesses in Nigeria to move to enviable levels by helping them reach their clients through its enormous nationwide data base of real phone numbers and email addresses. It has hundreds of organizations as its clients including multinational companies like Guarantee Trust Bank, PZ Cussons, MTN, Chivita, among others. He is also the Founder and President of Infinity Foundation and Stephen Akintayo Foundation, an indigenous non-governmental organization that assists orphans and vulnerable children as well as mentor young minds. The foundation has assisted over 2,000 orphans and vulnerable children and has also partnered with 22 orphanage homes in the country. By December 2015 Infinity Foundation is starting Mercy Orphanage to care for victims of Boko haram attacks in the Northern part of Nigeria.

Stephen Akintayo Foundation focus on Financial Grants with Initial grant of 10,000,000 to 20 entrepreneur in 2015 plan to grow that to 500Million annual grant by the 5th year. Projects like Upgrade Conference and The Serial Entrepreneur Conference with thousands of attendee who benefit from the high value knowledge from exceptional speakers and consultants.

Stephen, popularly called Pastor Stephen is also the founder of Omonaija, an online radio station in Lagos currently streaming for 24 hours daily with the capacity to reach every country of the world.

He is the founder and Director of Digital Marketing School Nigeria. Africa's leading Digital Marketing School issuing diploma certificates with robust training curriculum in Digital Marketing, Tele Marketing and Neuro Marketing. He is an Author of several published books including Turning Your Mess To Message, Soul Mate, Survival Instincts and Mobile Millionaire

Stephen is a media personality in the Television, Radio and Print media. He is currently anchoring a programme on Radio Continental, tagged CEO Mentorship with Stephen Akintayo, and A TV Show coming airing in the last quarter of 2015 as well as currently running a weekly column in some of Nigeria's national papers, including The Nation Newspaper and The Union Newspapers. He is also a social media guru.

His mentorship platform has helped thousands of people including graduates and undergraduates in the area of business as well as in relationships. Stephen strongly believes young Nigerians with the passion for entrepreneurship can cause a business revolution in Nigeria and the world at large. Stephen Akintayo is currently running Masters In Digital Marketing and MBA in Netherlands.

He is a trained Digital Marketing Consultant By The Digital Marketing Institute and Harvard University. He is also a trained Coach by The Coaching Academy UK. He has several other professional training inside and outside Nigeria.

He is First Degree is in Microbiology from Olabisi Onabanjo University, a member of Institute of Strategic Management. He is an ordained Pastor with Living Faith Church Worldwide and he is happily married and blessed with Two Sons; Divine Surprises and Future.

To invite Stephen Akintayo for a speaking engagement kindly email **invite@stephenakintayo.com** or call: 08188220066.

Website:	http://stephenakintayo.com
	http://gtext.com.ng
	http://infinityfoundation.org.ng
Email:	info@stephenakintayo.com
	Stephenakintayo@gmail.com

CHAPTER ONE
THE POWER OF THE MOBILE

CHAPTER ONE

THE POWER OF THE MOBILE

A billion mobile subscribers were added in the last 4 years to leave the total users of mobile communications standing at 3.2 billion – almost half the world's population. Mobile technology (and devices) form a critical part of innovation technology, which in the 21st century is referred to as the "Third Platform". Mobile (or Smart) devices — portable tools that connect to the internet — have become a part of our lives. In the last quarter of 2010, sales of smart phones outpaced those of PCs for the first time, according to data from IDC. By 2014, more smart devices could be used to access the internet than traditional computers. There is an obvious move to an increasingly mobile world, and this is creating new players and new opportunities for a variety of industries.

According to IDC: Over the next four years, the number of people accessing the Internet through PCs will shrink by 15 million as the number of mobile users increases by 91 million. In 2015, there will be more consumers accessing the Internet through mobile devices than through PCs. The picture below gives a graph of the total world population connected to a mobile device.

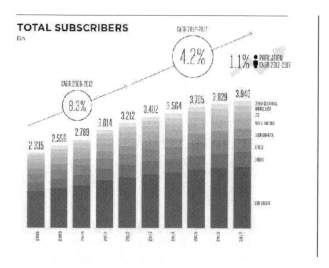

Source: GSMA Wireless Intelligence

Emerging markets will also create plenty of opportunities related to smart technology, and this will not be limited to for-profit enterprises.

As smart devices become increasingly accepted, companies are also moving into adjacent markets to exploit new revenue models such as mobile commerce (m-commerce) and mobile payment systems. It should be noted that a number of data and tech giants are already jockeying for position in this area.

This growth is mirrored by strong mobile connections growth, to almost 7 billion connections in 2012, as many consumers have multiple devices or use multiple SIMs to access the best tariffs, while firms in many industry sectors roll out M2M applications to boost their own productivity and tap into new markets. Despite challenging economic headwinds in many regions, the market is expected to grow even more strongly on the dimension of connections over the next five years, with 3 billion additional connections expected to be added between 2012 and 2017, a growth rate of 7.6% p.a.

The figure below gives the number of active mobile networks subscribers all over the world:

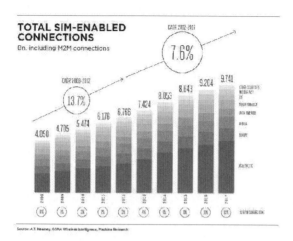

Benefits of the Mobile Technology

1.. Reductions in cost of managing health information systems. In Kenya, for example, mobile phones are being used to collect data and report on disease-specific issues from more than 175 health centers serving over 1 million people. This technology has reduced the cost of the country's health information system by 25% and cut the time needed to report the information from four weeks to one week.

2. Mobile devices give individuals constant access to information, regardless of location, anytime.

3. They also provide broad social benefits such as remote access to education and health care information.

4. Smart devices help to trigger an information explosion that blurs the traditional boundaries within, and across, industries.

5. It empowers consumers and provides new and huge opportunities for businesses. It should be noted, however, that mobile technology also poses a disruptive threat to businesses and individuals that cannot flow with this trend.

Opportunities Provided By Mobile Technology

1. Given this dynamism, it is no surprise that the mobile industry makes a substantial economic contribution, with mobile operators alone expected to contribute 1.4% to global GDP in 2012 and their revenues expected to grow at a robust 2.3% p.a. to reach US$1.1 trillion by 2017. When the rest of the mobile ecosystem is included, total revenues are forecast by A.T. Kearney to reach US$2 trillion in 2017, which represents an annual growth rate of 4.7%. In support of the growth in capacity and the level of innovation, the mobile ecosystem will increase its level of annual capital expenditure by just under 4% per annum from 2012 to 2017 to US$238 billion.

2. Mobile infrastructure is now as important to a country's economy as its energy grid or transportation network – it is a key enabling infrastructure that drives and supports growth in the wider economy. The high level of investment is needed to continue the development of this infrastructure so that capacity can be built to meet the ever growing demand and so that new services can be launched which bring greater benefits to the wider economy.

3. The mobile industry has always made a significant contribution to public funding. By 2017, its contribution to public funding is projected to be US$550 billion – as a result of spectrum fees as well as direct and indirect taxes. It is important that this level of financial commitment should be structured in a fair and predictable manner, in order to protect growth and employment – for instance the industry already supports 8.5 million jobs today and growth is expected in emerging markets to create an additional 1.3 million jobs by 2017.

4. The mobile industry is working to support and protect citizens. From empowering women or protecting the vulnerable to helping responses to natural and man-made disasters, mobile phones have significant potential to change the lives of millions. But with any new technology comes new risks and the whole mobile ecosystem is collaborating to reduce risks to users such as handset theft, mobile fraud and breach of privacy. Mobile operators are also playing their part to work to reduce their impact on the environment and have the potential to make a net positive impact on greenhouse gas emissions – with the potential to enable emission savings in 2020 more than 11 times greater than their own anticipated mobile network emissions. Across all of these areas, the GSMA is leading the drive to implement innovative new services and implement robust protective measures through the sharing of best practices and by encouraging inter and intra-industry co-operation.

5. The Social Media. There are more than 680 million active monthly mobile users on Facebook, 120 million active monthly mobile users on Twitter, 40 million active monthly mobile users on Foursquare, and 46 million active monthly mobile users on LinkedIn.

6. Increased phone engagements. An average individual spends daily 25 minutes on the internet, 17 minutes on the social media, 16 minutes doing music, 14 minutes on games, 12 minutes on calls, 11 minutes on emails, 10 minutes on SMS, 9 minutes watching video, 9 minutes reading books, and 3 minutes taking or watching pictures, all with the advent of Smartphone combined with the widespread deployment of mobile broadband networks.

Facts about Mobile Devices

- Mobile device online dominance: Mobile devices are the new primary design point for end-user access

- Soaring sales rates in mobile devices: Mobile devices are the new primary design point for end-user access. Participating across the full spectrum of mobile devices (smart phones, mini tablets, full-size tablets, PC/tablet hybrids, etc.) and aligning with mobile platforms that win the battle for developers and apps is the essential recipe for end user device manufacturers; seeing these as wholly distinct markets is an obsolete vision.

- Mobile application (app) platforms: Mobile platforms that, by the end of 2013, fail to crack the 50% barrier of developers that are "very interested" in developing apps for them will be on a gradual track to demise

- The real "PC versus mobile device" battle is between PC software platforms (especially Windows) and the leading mobile device platforms (iOS and Android). And the market power of these competing platforms — iOS, Android, Windows, and other mobile software platforms — will depend completely on the ability of each platform to attract large numbers of app developers.

- Strategic customer communities: In 2013, the accelerating shift to the 3rd Platform (of which mobile devices is chief) will continue to raise the profile of key customer communities — communities that are driving industry growth and redefining the design points of successful offerings.

- Worldwide IT spending will grow by 5.7 % , thanks largely to mobile devices.

The picture below shows the interest of Mobile App Developers in major platforms:

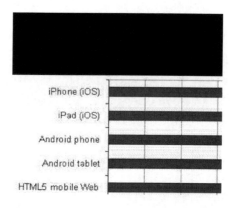

HINDRANCES TO BEING CONNECTED TO THE MOBILE WEB

There are still many adults and young people who would appreciate the social and economic benefits of mobile technology but are unable to access it, highlighting a huge opportunity for future growth and a challenge to all players in the industry ecosystem to expand the scope of products and services to tap this demand. Some of the easily identified challenges being:

- Expensive devices. In many developing countries, due to high poverty rate, some of the population considers these devices as too expensive.

- Expensive service plans. For some others, the service plans are simply too expensive.

- Poor mobile network. Network is usually better in the city centers. At the suburbs, there are simply no mobile networks to connect to.

- Content isn't available in the local language.

- Awareness of the value of internet is limited.

- Availability of power sources is limited.

- Networks can't support large amounts of data.

CHAPTER TWO
THE CONCEPT OF MOBILE MARKETING

CHAPTER TWO
THE CONCEPT OF MOBILE MARKETING

Mobile marketing is marketing on or with a mobile device, such as a Smartphone. Mobile marketing can provide customers with time and location sensitive, personalized information that promotes goods, services and ideas. Mobile (Digital) Marketing is a veritable tool both as a business and as business-promoting tool. Personally, I make over Five Hundred Thousand Naira (Nigerian currency) each month doing mobile marketing (alone). It should be noted that mobile marketing can be both a business, and can also be a business-promoting tool. This section seeks to explore some of the commonest mobile marketing platforms explored by the mobile millionaires (of which I am a player too).

Mobile ads: is a form of advertising via mobile (wireless) phones or other mobile devices. It is a subset of mobile marketing. Mobile advertising is targeted at mobile phones, a cost value that came estimably to a global total of 4.6 billion as of 2009. The 51 emergence of this form of advertising is so real that there is now a dedicated global awards ceremony organized every year by Visiongain. As mobile phones outnumber TV sets by over 3 to 1, and PC based internet users by over 4 to 1, and the total laptop and desktop PC population by nearly 5 to 1, advertisers in many markets have recently rushed to this media. In Spain 75% of mobile phone owners receive ads, in France 62% and in Japan 54%. More remarkably as mobile advertising matures, like in the most advanced markets, the user involvement also matures. In Japan today, already 44% of mobile phone owners click on ads they receive on their phones. Mobile advertising was worth 900 million dollars in Japan alone. According to the research firm Berg Insight the global mobile advertising market that was estimated to € 1 billion in 2008. Furthermore, Berg Insight forecasts the global mobile advertising market to grow at a compound annual growth rate of 43 percent to € 8.7 billion in 2014. It is my earnest hope and desire that such data will be made available for Africa, especially Nigeria, the most populous black nation, so that we can maximize this invaluable asset

Common Types of Mobile Ads
- Mobile Web Banner (top of page)
- Mobile Web Poster (bottom of page banner),
- SMS advertising (which has been estimated at over 90% of mobile marketing revenue worldwide).
- MMS advertising,
- Advertising within mobile games
- Advertising within mobile videos, and during mobile TV receipt,
- Full-screen interstitials, which appear while a requested item of mobile content or mobile web page is loading up, and
- Audio advertisements that can take the form of a jingle before a voicemail recording, or an audio recording played while interacting with a telephone-based service such as movie ticketing or directory assistance.

How to Measure the Effectiveness of Mobile Ad Campaigns

The effectiveness of a mobile media ad campaign can be measured in a variety of ways. The main measurements are

- Views (Cost per Impression): the number of times target customers view the ad campaign.
- Click-through (Cost Per Click): this involves the target clicking on the ad; he may or may not make a buying decision eventually
- Click-to-call rates: this involves the target clicking the ad, and eventually making a decision either to call for more information, or actually making a buying decision.

- Cost per Install (CPI) where there the pricing model is based on the user installing an App on their mobile phone. CPI Mobile Advertising Networks work either as incent or non-incent. In the incent model the user is given virtual points or rewards to install the game or App.

SMS Marketing: this is marketing that is done through mobile phones' SMS (Short Message Service). It became increasingly popular in the early 2000s in Europe and other parts of the world when businesses started to collect mobile phone numbers and send off wanted (or unwanted) content.
On average, SMS messages are read within four minutes, making them highly convertible.

Mobile Commerce: The phrase mobile commerce was originally coined in 1997 to mean "the delivery of electronic commerce capabilities directly into the consumer's hand, anywhere, via wireless technology." It is the use of wireless handheld devices such as cellular phones and laptops to conduct commercial transactions online. Mobile commerce transactions continue to grow, and the term includes the purchase and sale of a wide range of goods and services, online banking, bill payment, information delivery and so on. It is also known as m-commerce.

According to BI Intelligence in January 2013, 29% of mobile users have now made a purchase with their phones. Wal-Mart estimated that 40% of all visits to their internet shopping site in
December 2012 was from a mobile device. Bank of America predicts $67.1 billion in purchases will be made from mobile devices by European and U.S. shoppers in 2015. Mobile retailers
in UK alone are expected to increase revenues up to 31% in FY 2013–14.

Common Products and Services Available

The commonest products and services available via the mobile commerce include (but not limited to) the following:

- **Mobile Money Transfer**: this generally refers to payment services operated under financial regulations and performed from or via a mobile device. Instead of paying with cash, cheque, or credit cards, a consumer can use a mobile phone to pay for a wide range of services and digital or hard goods. Common mobile payment platforms in Nigeria are: MyPesa, Paga, M-Teller, M-Naira, VTN, M-Wallet, Monitise, Access Mobile, Enterprise Mobile, Diamond Mobile, SwipeMax Wallet, Breeze Nigeria, Sterling Mobile, Wema Verve, EaZyMoney, QuickTeller, etc.

- **Mobile ATM**: The mobile ATM device easily attaches to most Smartphones and dispenses money instantly and effortlessly– forever ending your search for the nearest bank or ATM. Just type in your personal pin code on your cell phone and access all your cash from the palm of your hand.

- **Mobile ticketing**: This is the process whereby customers can order, pay for, obtain and validate tickets from any location and at any time using mobile phones or other mobile handsets. Mobile tickets reduce the production and distribution costs connected with traditional paperbased ticketing channels and increase customer convenience by providing new and simple ways to purchase tickets.

- **Mobile vouchers, coupons and loyalty cards**: Mobile ticketing technology can also be used for the distribution of vouchers, coupons, and loyalty cards. These items are represented by a virtual token that is sent to the mobile phone. A customer presenting a mobile phone with one of these tokens at the point of sale receives the same benefits as if they had the traditional token. Stores may send coupons to customers using location-based services to determine when the customer is nearby.

- **Content purchase and delivery**: Currently, mobile content purchase and delivery mainly consists of the sale of ringtones, wallpapers, and games for mobile phones. The convergence of mobile phones, portable audio players, and video players into a single device is increasing the purchase and delivery of full-length music tracks and video. The download speeds available with 4G networks make it possible to buy a movie on a mobile device in a couple of seconds.

- **Location-based services**: Location-based services (LBS) are a general class of computer program-level services that use location data to control features. As such LBS is an information service and has a number of uses in social networking today as an entertainment service, which is accessible with mobile devices through the mobile network and which uses information on the geographical position of the mobile device. This has become more and more important with the expansion of the Smartphone and tablet markets as well. The location of the mobile phone user is an important piece of information used during mobile commerce or mcommerce transactions. Knowing the location of the user allows for location-based services such as:
· Local discount offers
· Local weather
· Tracking and monitoring of people

- **Information services**: A wide variety of information services can be delivered to mobile phone users in much the same way as it is delivered to PCs. These services include:
o News
o Stock quotes
o Sports scores
o Financial records
o Traffic reporting
Customized traffic information, based on a user's actual travel patterns, can be sent to a mobile device. This customized data is more useful than a generic traffic report broadcast, but was impractical before the invention of modern mobile devices due to the bandwidth requirements.

- **Mobile Banking**: Banks and other financial institutions use mobile commerce to allow their customers to access account information and make transactions, such as purchasing stocks, remitting money. This service is often referred to as Mobile Banking, or M-Banking.

- **Mobile brokerage**: Stock market services offered via mobile devices have also become more popular and are known as Mobile Brokerage. They allow the subscriber to react to market developments in a timely fashion and irrespective of their physical location.

- **Auctions**: Over the past three years mobile reverse auction solutions have grown in popularity. Unlike traditional auctions, the reverse auction (or low-bid auction) bills the consumer's phone each time they place a bid. Many mobile SMS commerce solutions rely on a one-time purchase or one-time subscription; however, reverse auctions offer a high return for the mobile vendor as they require the consumer to make multiple transactions over a long period of time.

- **Mobile browsing**: Using a mobile browser — a World Wide Web browser on a mobile device customers can shop online without having to be at their personal computer.

- **Mobile purchase**: Catalog merchants can accept orders from customers electronically, via the customer's mobile device. In some cases, the merchant may even deliver the catalog electronically, rather than mailing a paper catalog to the customer. Some merchants provide mobile websites that are customized for the smaller screen and limited user interface of a mobile device.

- In-application mobile phone payments: Payments can be made directly inside of an application running on a popular Smartphone operating system, such as Google Android. Analyst firm Gartner expects in-application purchases to drive 41 percent of app store (also referred to as mobile software distribution platforms) revenue in 2016. In-app purchases can be used to buy virtual goods, new and other mobile content and is ultimately billed by mobile carriers rather than the app stores themselves. Ericsson's IPX mobile commerce system is used by 120 mobile carriers to offer payment options such as try before-you-buy, rentals and subscriptions.

CHAPTER THREE
BULK SMS EXPLAINED – THE MONEY MAKER

CHAPTER THREE
BULK SMS EXPLAINED – THE MONEY MAKER

Bulk SMS Statistics

65% of people send business SMS from their mobile phone on a daily basis

50% of people receive business SMS from a member of their team/ department/ office

47% of people receive business SMS from a colleague

36% receive business SMS from a customer or client

71% of people use business SMS for setting 'reminders of meetings and appointments'

18% of people used business SMS for 'notifying people about new meetings and appointments'

14% of people use business SMS for 'chasing up suppliers or new orders'

A third of respondents (36%) said that they receive SMS from a customer or client

11% of people receive SMS from suppliers

I'll leave you to draw your conclusions. Good luck in your bulk SMS marketing campaigns.

A. Mobile Marketing

Although mobile phones have been around for some time now, smartphones are really beginning to take off. In 2013, over one billion smartphones were shipped in a year for the first time ever. Despite this, many firms aren't taking advantage of SMS marketing. Here are some impressive statistics that will show you just how big an impact SMS marketing can have on your business.

1) Smartphone ownership: Some 5.1 billion out of the 6.8 billion people on Earth own a mobile phone - that's a huge market.

2) Fast response times: On average, it takes 90 minutes for someone to respond to an email, but of course some people might takes hours or days to even look at it. With text messages, the average response time is 90 seconds.

3) Open rates: Around 98 per cent of all SMS messages are opened, but only 20 per cent of emails are looked at. It's a similar story for social media too; only 12 per cent of Facebook posts and 29 per cent of tweets are read.

At GILEADBALM GROUP, we specialize in driving business growth through mobile marketing, and other strategic marketing initiatives. We want to make it simple for small & medium sized businesses, organizations and individuals to leverage the various social platforms to grow business leads, drive web traffic, expand market reach and convert more sales, faster. Our certified mobile marketing experts have created a proven mobile marketing workflow to do just that – build brand awareness, foster engagement and gain more clients.

Plan A is To Own Your Website

Build an SMS website with an appealing design and coded to send sms, allocate credits, save user details and to do other tasks associated with the bulk sms business. Unless you are a programmer, you will need to engage the service of a programmer to do this.
Register a domain name for your website.
Acquire a web hosting plan. It is through web hosting that your website can be accessed from anywhere in the world.
Acquire an SMS Gateway Account from an SMS gateway provider. Like www.gileadsms.com Consider rates charged by the provider so that you will be able to offer competitive rates in turn to your customers.

Determine how much you will charge your customers. To do this, take into account how much you are paying for the units acquired from your provider.

Link your SMS website to the SMS gateway provider with whom you have created an account.

Fund your account with the service provider.

Market your service.

Provide bulk sms for your customers.

Ensure customer satisfaction through timely delivery, high delivery rate and good customer relations.

Get GSM Datbase of your local country and add it to your site so that client can send targeted campaigns as most of your client that are SMEs will need sms for the purpose of marketing but please be sure to read your countries law and make sure it is still legal. In A Country like Nigeria, It is still legal because no anti-spam law yet.

B. Operating from a provider's platform

Instead of building your own website to be linked to a provider's site, you can simply avail yourself of the use of a platform provided free by a provider. This way, you avoid the expense of building an SMS platform. Here are the steps to follow to go this way.

Register with a bulk sms provider. Since you will be using the provider's platform, care should be taken to select a provider whose platform offers a package that is as comprehensive as you can get. Also consider rates charged by the provider so that you will be able to offer competitive rates in turn to your customers.

Determine how much you will charge your customers. To do this, take into account how much you are paying for the units acquired from your provider.

As soon as you have registered, the provider will provide you with a portal or platform from which you will operate.

Get your portal customized. This is done for you by the provider, using information that you provide.

Fund your account with the provider.

Market your business.

Provide bulk sms to your customers.

Ensure customer satisfaction through timely delivery, high delivery rate and good customer relations.

Whichever method you choose, the basic requirement for the business remains the computer and internet connection. Very little space is required but marketing will need a lot of time and attention. Entry into the business is easy and accounting for profit is straightforward.

WHO NEEDS BULK SMS?

There are many categories of organizations that need Bulk SMS, these include:

- Churches
- Mosques
- Schools
- Supermarkets
- Companies
- Hospitals
- Clubs
- University Departments
- Wedding Announcement & Invitations
- Events and Meeting
- Birthday Announcement and Invitation
- Special Seasons Greetings
- Political Campaign and Awareness
- Corporate Bodies
- Government Establishments
- Associations and so on.

These organizations need Bulk SMS service for different purposes, such as:
- Advertising Campaign
- Disseminating Information
- Invitation
- Notice of Meeting
- Statement of Account and many more…

Fact – 1

Let's use Nigeria as a case study but note that this business is global and we can help you set your bulk sms business anywhere in the world.

- Do you know that 1 billion SMS messages are sent by Nigerians Per Year? Now what this means is that N15 billion Naira is being shared annually between a small group of smart & average Nigerians?!

Fact – 2

• Do you know that many Nigerians are quickly moving from phone messaging to bulk SMS messaging especially small organizations because it's cheaper to use online bulk SMS?

Fact – 3

• Have you heard about recent developments in the telecom industry? That more than 40bmillion Nigerians now use GSM phones and the number keeps increasing daily?

See This Simple Calculation:

Maybe you visited a Company as a Bulk SMS Client, Meaning you want to handle their Bulk SMS Work, and they accept your Proposal, You can Charge them for N1:50-N1"2:50 per SMS, if you have targeted GSM phone number database you can even charge more depending on the country you live in. Let me do a Little Calculation

Bulk SMS Company sells SMS for you @ the Rate of N1-N2Each! (Depending on the amount of unit purchased)

———————————

You Charge the Company who need your SMS Service for N3-N4Each Particularly if you have classified numbers that can help them send targeted messages.

———————————

Then they Decided to Send to 50,000 Customers, Then you have to
Purchase 50,000 SMS units.
ANSWER:

You have Automatically Make:
50,000 SMS * N2= N100,000, Then you can now Deduct the Money you Used in
Purchasing the SMS which is 50,000 * N1 or N1:4= N50,000 or N70,000. Rest is your profit.

How To Build A Successful Bulk SMS Business With A System In Place

Bulk SMS business is one of the great ways of making cool money. In this article, you will learn secrets about bulk SMS business and why it is so easy to make cool money from bulk SMS business.

What I like most in bulk SMS business is that I can be in my room and still be making money. I can travel to anywhere and still be making money. I make money every day from my bulk SMS business. It is amazing. You too can make your bulk SMS business real fun as you make more money using the secrets revealed in this article. I will explain all the information you need to define your own bulk SMS business model and make profit with ease. If you are just about to start bulk SMS business, I encourage you to do so quickly before it will be too late. This article contains all the secrets that highly successful bulk SMS website owners know and make use of.

The cost of Bulk SMS website designing is usually between N9,500 and N35,000. You too can also have other bulk SMS website owners to resale your bulk SMS. The more bulk SMS website owners you build, the more bulk SMS you sell. These bulk SMS website owners would bring in more customers indirectly into your business and save you huge cost on advertisement.

In addition, you can use bulk SMS resell model to expand and get more customers faster than you can imagine. For instance, if you have 100 bulk SMS users on your own bulk SMS website and I have just 10 bulk SMS website owners to resell my bulk SMS, I have advantage of making more money regularly than you. As a bulk SMS website owner, you have to focus more on building bulk SMS website owners than building end users.

How to Handle Competition in Bulk SMS Business

There are competitors out there but it shouldn't be a problem if you know what to do. The major challenge of most of the bulk SMS website owners is lack of a system. You must know that anything that works is based on a system. You must have your own system of running bulk SMS business. I have developed and used one of the best bulk SMS systems in the world. Your system defines your customers, prices and marketing. When you have all these setup like an autopilot, then competitors will be afraid of you. You can crush any competitor because many of them just jumped into bulk SMS business without a defined system.

A system is your understanding of how bulk SMS business should work best. Your system should show you who your customers are, where they are and how best to serve them. With a well-defined system, you will not have a competitor to crush.

A good system would have worked successfully before others may consider copying it. The time they are waiting to see you succeed gives you enough time to build your business to withstand competition. If your competitors are those who are just coming up, then you really don't have competitors.

Sometimes you may even think that your competitors are doing very well more than you. I have heard most bulk SMS website owners complained about some other bulk SMS website owners' services. They said their prices are better and their websites are easier to use. How did you know? Is it because you used it and it is easy for you? Does it mean that others find it easier too? Before you start condemning your own service, try and find out what the competitors are saying about you. Most of the so called competitors are new in bulk SMS business and probably don't really know how the business works. Don't ever make a mistake of being desperate. Rather, stick to your bulk SMS business model.

Most bulk SMS website owners would get frustrated earlier in the business and that is why I'm not afraid of new competitors. They are new and they don't know the best way to enter the market. They are only puffed up by their seemingly good calculations. I can imagine them calculating that if they sell 1 million bulk SMS this month at N1 profit, it would be N1 million profit. It sounds great but the question is how are they going to do it. How are you going to sell 1 million bulk SMS this month? I know they don't have answers to those questions. Therefore, I don't bother about them when they get frustrated and start selling at a lower price.

How To Hit 100,000 Bulk SMS Monthly As New Bulk SMS Website Owner

If you are new in bulk SMS business, you could be scared to think of selling 100,000 bulk SMS within the first 3 months but it is possible. Selling 100,000 bulk SMS within the first 3 months is even more possible now than you think. You can use my strategy that helped me when I started as a beginner. I first had a simple target of selling 10,000 bulk SMS. This may seem small if you consider the profit you can make from it but I want you to forget about profit. Bulk SMS profit is volume driven. To get it, you must have volume. It would be a wrong move to start thinking of profit. I will advise you to think of volume. If you have the volume you will necessarily make profit by default.

After you have set your target of 10,000 bulk SMS, consider the bulk SMS business model you are using. I recommend direct marketing to friends, church members, family members, colleagues at work, etc. Where to find a good list of target customers to send text message advert about your bulk SMS business to is in your own phone-book. This should be your first advert. Wait for response. Try to send a reminder text message after 3 days.

Keep record of those people you have told about your new bulk SMS business. The more the number grows, the more responses you will get. Your first 10,000 bulk SMSsales should happen in your first week or month. Your target for the second month should be 100,000 bulk SMS. This time, try other advertisement sources like newspaper. I recommend adverts in papers that have your target customers. I hope you developed your own system. This is when to use your own system. Let it guide you and make you remain focused.

If you use your business model strictly, you will hit the 100,000 bulk SMS sales in your second month. If you don't make it in the second month, no problem. Do it again next month. It means you are closer to it. Also address your customers complaints that are real but don't get carried away by complaints on cheap prices from competitors. Customers always want cheap prices, so they would always complain about price. Stay within the average price range, but I strongly recommend you consult your bulk SMS business model.

What successful bulk SMS Millionaires are doing that you should know and do:

Many bulk SMS businesses never make it to the profit lane because of what I want to share with you here. Successful bulk SMS Millionaires define who their target customers are and they just approach them. In the bulk SMS market, there are different types of customers. Those who use bulk SMS and those who use bulk SMS service. Those who use bulk SMS are low volume customers. They just use bulk SMS because it's cheaper than their regular phone rates. They are the type of customers that use 100 to 500 bulk SMS volume monthly. If you focus on this kind of customers, you would require huge customer base to sell good volume. This kind of customer has its own advantages. The other category is those who use bulk SMS service, and they are those whose business and/or daily/weekly/monthly activities depend on bulk SMS service to thrive. Examples are businesses/organizations, churches, mosques, banks, etc.

Imagine if your customers are those whose businesses depend on bulk SMS service to thrive! You will have regular customers that will keep your bulk SMS business profitable at all times. If your customers are those whose businesses do not depend on bulk SMS to thrive, the opposite is the case. It's better you choose who your customers are carefully. Stick to your business model and you will begin to see results.

Let's even be more practical. If you have customers that buy just 1,000 bulk SMS each, you will need 1,000 customers to sell 1 million bulk SMS. If your customers are those who buy 5,000 bulk SMS each, you will need 200 customers to sell 1 million bulk SMS. If your customers are those who buy 10,000 bulk SMS each, you will need 100 customers to sell 1 million bulk SMS. If your customers are those who buy 100,000 bulk SMS each, you will need 10 customers to sell 1 million bulk SMS. If your customers are those who buy 1 million bulk SMS each, you will need just 1 customer to sell 1 million bulk SMS.

Imagine you have 100 bulk SMS customers that buy at least 5,000 bulk SMS on average per month. That's a total of 500,000 bulk SMS monthly. If a competitor who has not sold its first 10,000 bulk SMS comes with lower price, what would you do? The answer on many people's mind will be to lower theirs too, not so.

What you need to do is to look at the market and see if there is a need for such lower price and what is the resultant effects on your bottom line. If it's not feasible, simply ignore it and rather offer bonuses to your customers to cushion any effect. Remember, you are most likely to be the one that knows that such competitor exist. Your customers might not even know.

Now my question is which customer would you rather have? Then go for them. Don't bother about those who are asking for this and that. Just know who your customers are and what they want.

Three (3) important bulk SMS marketing SECRETS that has been TESTED & PROVEN by highly successful bulk SMS resellers that will explode you into the millionaires circle in just 120 days:

These 3 secrets are target customers, the right platform and the right massage. If you want your bulk SMS business to be profitable, you must understand these 3 marketing secrets and use them well.
1) Who are they?
2) Where can you find them?
3) What do they want to hear?

If you put these three things together you will always win in bulk SMS business. It's your target customers that would define the platform and the message. Don't just advertise on any platform or draft advert message without a knowledge of the target customer.

If your customers are students for example, it helps you to know where they are, in school and what they would like to hear. Students would always like to hear the words cheap and free. So, for you to successfully market to them you must take your business to their schools, and your message should either have cheap or free, or both. That means you would either have something free to offer them or you create a cheap offer to win their patronage.

Bulk SMS RESEARCH STUDY:

As is in every business, there is crucial need for feasibility study so also BULK SMS Business In Nigeria is not left but, because during this study you get to know your target market, cheap and reliable companies with best reselling plan and price tag and also it teaches you to know how to tag price your units to beat the whooping competition. With conception we now believe that the first steps to take when venturing into the bulk SMS in Nigeria is to do in-depth research and study into the business, see into what others already in the business are doing and how there do it. Absolutely when you do this, there would be no cause for alarm-sparked mistakes because cautions will be in play.

FINANCIAL PLAN FOR THE BUSINESS:

Back to those days when bulk SMS business was in its early bird stage, starting a bulk SMS business was a huge break up both in capital and time, but gone are those days, now a barely N10,000 can start a bulk SMS business as a middle man and N22, 500 as a owner of the business with a designed and installed website, that you will get from your reselling or providing company. No doubt, bulk SMS business is now for the low capital and technological entrepreneur. Imagine, you paid just N22, 500 you get well designed website, hosted and branded, that means you start your business in a matter of one day and start making money the same or next day, it's simply time saving business.

But if your bulk SMS business website is up and running, how would you make sales is the question that should prick your mind, which mean, you need to get 1-5 customers daily to make it in the business. What comes to your mind is how you get these customers that would give you money: up next is what I will be exposing to you, so let's go there.

HOW TO GET 1-5 CUSTOMERS DAILY:

Put in Your marketing strategy, for me, I took sometime to stumble and summarize the long procedures about this business into a simplify and easy to understand format. For your bulk SMS to make a surprising break in profit you need to strategize your working style of marketing, tactics that works for you, that uprightly means you should test all the marketing pin-point and then take note effectively which and which work out well. Now, here are some of the marketing strategies:

WORD OF MOUTH:

Almost every business owner is a marketer and every marketer knows how to use their tongue in a sweetly and thrilling manner, they are always sweet tongued. Say the words with style, spread the word (tell everyone about the business and that you are a business man or woman). Know when to say it so as to make it memorable to the person right under your mouth. Am done with this.

CLASSIFIED ADVERTISEMENT:

So far as Nigeria is concerned, newspaper are most resourceful point, so placing a tiny call to action and hyper-super attractive headlines classified advert is a sure killing marketing strategy. Here is a sample on how your advert should look like " REVEALED!!!

Cheap bulk SMS 70Kobo PER SMS Unit. 2Unit equal 1page sms. Meaning 1:40kobo to 1sms.
Instant delivery, instant feedback to sender with 50% discount.
www.yourwebsite.com. "

If you can come up with such an advert and place it on most popular Nigerian newspapers, certainly your sales will be skyrocket. Prepare a plan, it may be weekly, monthly or daily and this will guide you through, if you should advertise monthly, weekly or daily.

Blogging:

I have come to notice that blogging is the greatest tools for brand and fans or community building yes, it work's for almost every type of business is not left, blogging expose and show your real self and business actions to your loyal readers that's why its called blog marketing. you should start a blog as a sister site to your bulk SMS website, teach and explain tutorial and, secret about SMS and bulk SMS in general with your marketing target – your bulk sms business website. I rest my case here concerning blogging (certainly most to come about this in my future post), up next still on marketing is.

Online advertising classified:
Basically before now, it has never been so easy to attract huge traffic to a site with just a simple of five lines sentence classified advert, but now is so easy and you should take advantage of the technology, how?, you must have heard about facebook ads, Google adwords, addynamos, and adbrite etc. My advice is specifically go for facebook advertising because you can first use their insight tools to test the ads and take account of its likeness and click-through rate. Open ads account, set a budget (say about 30 dollars a day), design a compelling and call to action ads, and finally test the ads, if it works good, play around facebook with your ads; that enough for social media and online advertising.

Furthermore, why you should start this bulk sms business now if you have found the idea keening; reasons.

Bulk sms Business is very profitable: 50% to 200% profit from a business is what I called profitable business and bulk sms business guaranteed this rate of profit percentage. This depends on how much you buy the sms that you will resell. Many people are getting into this business because of its profit potential and only very few business in Nigeria have this similar trait.

Bulk sms business is not expensive to start: capital is the biggest complain every young entrepreneur always have, but with the bulk sms business, there is no need for so much capital, your complains is 50% off, you start with as low as N25,000 (price of round table with few drinks and pepper soup). You don't need to get a loan either from anyone or bank that makes you going without fingers.

Bulk sms business is dirty easy to set-up: in most case, you have to pay your money into the provider accounts and then go and take a sleep, while your provider design, installed, host and test-run the new business for before handling over it to you . No complex, technical thing or skills, yes technical can help you if you have them but is not required.

Bulk sms Business does not required an office and is not time consuming type of business. You can run the business from the comfort of your home, in cyber café or anywhere as part-time, fulltime or combine together. It is automated after setting it up (which in most case is done by your provider, all you have to do is to get customers. You can run it only on weekends, as a student and full time day-job-worker – the leverage is much.

Finally, the growing demand is high:
Think about this the more and more people use bulk sms service to broadcast, announce and invite friends to their wedding, church services, ceremonies and the almighty bulk sms advertising – big business and small business use it and are looking for it. Let your sms units be cheap, fast in delivery and automated, then the money is yours. Conclusively, I rest my case about bulk sms business in Nigeria, may be in the future if there is demand for more details I will re-visit it again.

Challenges of Bulk SMS Business
1. **Technical problems** – I did have some Joomla issues like my website not sending registration emails to new users, my bulk SMS component crashing, losing my database, hacking, security issues, and all of that. But the moment I hired programmers, it's no big deal to me anymore.

2. **Online payment problems** – Forget about the lip-service talk of cashless policy by our government, they are not providing the infrastructure needed to support it like it is obtainable in other countries.

My customers find it difficult to pay with their ATM card online even though I've set everything right. Sometimes, their bank account is debited when the payment didn't go through. And when that happens, they naturally swear never to use it again. So, payment processing is one hell of a big problem. How do you go to the bank to pay N500?

3. Blocked sender ID and non-delivery of SMS by Nigerian GSM operators – Occasionally, the GSM networks in Nigeria block the sender ID from being displayed, instead they will display just some random numbers for all SMS emanating from the Web.

SMS from being delivered. And my providers won't refund the thousands of SMS that were not delivered nor will they resend it. And if messages are not delivered, then where is the business?

And what will you do to an MTN or a GLO that will hurt them? They are big companies that can afford to do as they like to Nigerian bulk SMS resellers and still turn in millions of profits. They don't care about you.

4. **Scammers activities** – Ever received an SMS telling you that you've won N1 billion from an ongoing MTN or Lacasera promo? That's what I'm talking about. That's one of the reasons our GSM networks filter bulk SMS from the Web because half of it is SPAM. And it's costing these GSM companies millions of naira to deliver millions of SMS every day from the Internet.

More so, I don't want people to use my bulk SMS platform to scam people, so I usually delete them manually. I have a system in place to alert me when a scam message is sent. But it's costing me in time and resources.

5. **Delivery Issue** - If your bulk sms portal is not delivering sms at all or having late delivery issue, then you are in for a serious business wreckage. It's better to have less attractive and unresponsive bulk sms website than to use premium joomla template to build your website, but your clients keep experiencing delivery issues. Pls avoid delivery problem at all cost. Go for premium route and reject economical route like cancer.

6. **Not Buying in Large Quantity** - This is one of the secrets of success in bulk sms business. Try to raise some fund, ask your friends or family members to invest in your bulk sms business. If you buy sms in large quantity, you will get it cheaper and then sell at a competitive price and recover your capital and income immediately. I remember asking clickatell.com for sms pricing. They told me #2/sms. That was outrageous because it will be hard to resell. I started negotiating with them and they said they will offer me at 80kobo/sms but that I must buy upto 5 million sms, which is 4 million Naira.

I am not saying you must raise millions of Naira if you must make it in this business. All am saying is that you should find a way to buy sms in large quantity and get sms cheaper.

7. **Using One Sms Gateway** - I personally use two very strong sms providers in the world. None of them ever failed me, but I insist on using two gateways to prevent any outage that will affect my business negatively. One of my providers don't use online payment and so for me to pay them, I usually buy dollars, pay into my domiciliary account and then wire to their bank account. It takes 2 to 3 days for the transfer to clear. So while waiting for the transfer to clear, I will be using the other company's API.

8. **Not Doing Enough Marketing** - Your bulk sms business can never move forward if you don't actively market it. You must advertise. I don't joke with adverts. I constantly look out for opportunities to leverage and push my bulk sms business to the people. Make out time and google about cheap business marketing strategies and implement some of them.

9. **Becoming Casual with it** - Your bulk sms business is a business. Don't make it a casual source of income. Get a good logo. Print a complimentary card. Have a brand colour. Use 'we' and 'our' anytime you are talking about your bulk sms. Bulk sms business is a serious business. Go register a business name at CAC. Get a corporate account in at least two banks. Get a dedicated phone number for the business.

There are challenges in bulk sms business, and so WHAT? Every business has it own challenge and this book has provided solutions to all of these and you too can do your research to help yourself .I tell people that the more the challenges, the more the Opportunity potentials. The more risk, the more profit. Hence this is one business that can change your fortune anywhere in the world, you can run a successful 7Figure bulk sms business.

The Principles of the Mobile Millionaires

There are quite some principles that the mobile millionaires adopt. Knowing and practicing these principles has made me a big player in the mobile world. These principles include:

The Principle of Simplicity: This principle states that whatever you do on the internet, especially with mobile technology, it must be easy enough for even the least educated person to use.

Whenever I consider this principle, I think of www.google.com. The search engine is so easy that every Jack can use it. The Principle of Complexity: This principle is also called The Power of More. It states that you must be willing to offer more than the users deserve. Do not hold back anything from your users. Whatever it is you are offering – a product, service (as discussed in previous chapters), or an app, give your intended users more.

The Principle of Niche: This principle states that every offering must have a clearly marked out audience in view. Do not try to be everything to everybody. Do not try to be a jack-of-all-trades. If, for example, you are composing a message for email marketing, have a target in view. Not everyone will be interested in your offering – no matter how appealing. But you can tailor your message to get the attention of your intended audience. If you don't, your message will be too watery to attract any response.

The Principle of Customer-Centredness: This principle states that you must work with your target customer or client in mind. It is not about what you like; it is about what your customer needs. No matter how bright your idea, no matter how brilliant your offering, if it does not meet the needs and aspirations of your clients, it will all be a waste. While writing this book, I had a challenge with one of my web designers. He had this beautiful concept about a particular website he developed for me, the website was beautiful, the concept was great, he was so enthusiastic about it, but my clients all complained that they had difficulties navigating the website. The concept was great but it did not meet my needs. Do not try this with your clients.

The Principle of Cross-Platform: This principle states that all your offerings, especially websites and apps, must be designed to work on every available device. Take for example a website; design your websites bearing in mind that not everyone will view it from a PC (computer). You must then ensure that whatever the device, visitors to your website will not have any issues viewing and navigating through the pages.

The Principle of Creativity: This principle states that you must be different to make a difference. People are used to the ordinary, general, as usual offerings. Give your clients some reasons to think you are crazy. Do something unique and a bit out of this world. Wow your clients – while not breaking the other principles. They will respect you for your creativity and be willing to pay for your services – even more than you deserve.

The Principle of Visibility: This principle states that your positioning determines your profiting. You must know how to position yourself via marketing. No matter how great what you are offering, if people don't know about it, you are like a handsome guy winking at a lady in the dark: there will be no response. You are doing something but only you know about it.

General Guidelines for Joining the Mobile Millionaires

- Identify your strengths, and work on them. Mobile Millionaires don't do things they are not good at. They focus on their strengths and delegate their weaknesses.

- Have a plan. Don't just breeze in to the mobile (or internet) world aimlessly. If you don't have a roadmap, you will soon lose your way to others who know why they are here.

- Fail forward. You will make mistakes and do things that will not work – that sounds normal. But it is insanity to make the same mistake twice. Fail. Learn. And move forward. Be better today than you were yesterday.

- Be tenacious. Don't think people will just make way for you because you have a foolproof idea. You can take my word for this – you will be kicked and tossed around till you can stand your ground. You must then make up your mind that it is not over till you win. Winners are losers who refuse to quit.

- Be consistent. You must always be in the face of the people. Once you begin, hang in there till you win. Don't undulate. Don't be away today, back tomorrow. Be predictable (that you will always be there).

- Be compassionate. Be genuinely interested in meeting the needs of your clients. Stay through with them till you solve their problems. Compassion is the only guarantee for an expansion. The more compassionate you are, the more attracted and connected you are to your clients.

- Be prompt. Be swift. Let your service delivery be unbeatable – in terms of speed and quality. The faster your delivery; the more jobs you can take on; the more money you make. Riches answer to volume of businesses done within the shortest space possible.

One More thing – Over to you.
Start your own bulk sms business today and start pulling 50% to 200% profit daily. Remember, don't eat your future – start something now!
Please avoid words like "congrats", "promo (ting)", "promotion" "wow", "splash", "yellow", "congrats" e.t.c in your messages.

In compliance to the directives of NIGERIAN Communications Commission (NCC) that banned all Nigeria Telecom Operators from engaging in PROMOTIONS and other LOTTERIES, Our service providers has blacklist some words/contents as SPAM and as such, messages contained same words or sender ids will not be delivered.

Find below the lists of SPAM FILTERING RESRICTIONS from our service providers.

Please note that (SMS) messages that contained these words will not be delivered.

Kindly pass this message to your friends as well.

|===

String(Text)
Spam Type
(at)yahoo.com
Both
..M"T N N
Both
.M"T N N
Both
1.8.O
Both
180 .
Both
2013winner
Both
30yrstc@w.cn
Both
41Recharge
Both
4-1Recharge
Both
A!rtel
Both

AIRTEL
Both
ANGEL/MEZZY
Both
AWOOF!!!
Both
BankAccess
Both
BANKM3py
Both
Bumazek
Both
C0NGRATS!
Both
C0NGRTS
Both
C0NGRTS!Y0U
Both
cantv.net
Both
Cards-Zone
Both
Card-Zone
Both
ccoca@live.co.uk
Both
CLAIM
Both
claim@ukmobilelotto.com
Both
claims@ukmobilelotto.com

Both
CLAIMTODAY
Both
clarkejudith93@yahoo.com.hk
Both
Coca-Cola
Both
coinmac.net
Both
cokegrant2012@live.com
Both
CONGRAT
Both
Congrats!
Both
Congratulation
Both
CONGRATULATIONS
Both
CouplesOut
Both
cw10187@yahoo.com.hk
Both
Debeeray?s
Both
esqhill@w.cn
Both
ETISALAT
Both
Euro Casino
Both

Gl0lwinner
Both
GLO
Both
Glo Promo!
Both
GLOBACOM
Both
Glowinner
Both
gnetsms@live.com
Both
GreatNews!
Both
gsm promo
Both
HSEFELOSHIP
Both
http://www.permanenttsb-updates.org
Both
http://www.yellopins.com
Both
info@nmobiledraw.org
Both
infonmobiledraw.org
Both
ISAMS YABA
Both
Jemtrade
Both
L21

Both
LIVINGWORD
Both
lo1O10
Both
lottery
Both
lotto
Both
M "T N N.
Both
M .T N N?
Both
M T N
Both
M T"N-N
Both
M T"N-N.
Both
M T"N-NG
Both
M"T N N.
Both
M"T N?
Both
M"T N-N.
Both
M"T N-N.?
Both
M"TN N
Both

M"TN-N
Both
M.T`N
Both
M.T`N GN
Both
M.T`N-.NG
Both
Maitap.be
Both
MANSARD
Both
Megateq
Both
mlottery@usa.com
Both
mlttryusa@w.cn
Both
mobile promo
Both
MT N
Both
MTEL
Both
MTN
Both
M-T-N
Both
-M-T-N-
Both
MTN N

Both
MTN NG
Both
MTN-NG
Both
N0K
Both
N10M
Both
N2Million
Both
nexteldraws@live.co.uk
Both
NKMOBILE
Both
NOK
Both
Nokia inc.
Both
nokia promo
Both
nokia_40years2@live.com
Both
nokia40years@live.co.uk
Both
nokia40yrs@hotmail.com
Both
nokialondondept247@hotmail.co.uk
Both
NOKIA UK
Both

Both
NTM
Both
OPEDAYO
Both
ORLAJ
Both
P C EBUNILO
Both
PR1ZE=0FFER
Both
PRIMEGROCER
Both
ROSGLORIOSM
Both
Service180
Both
Spam
Both
StanbicIBTC
Both
SUPOL SEGUN
Both
SWFT
Both
SWIFT 4G
Both
Swift_NG
Both
SWIFTNG

Both
takko
Both
telsms@live.com
Both
topkids
Both
TOYOTA sweepstakes
Both
tsms04@live.com
Both
ULTIMATE
Both
usa.mlty@w.cn
Both
usa_mlty@w.cn
Both
vadia
Both
VISAFONE
Both
W0N
Both
W1N
Both
wbre@gala.net
Both
wbre2@gala.net
Both
WIN
Both

wo n
Both
won
Both
www.mtndays.org
Both
www.mymtnsim.com
Both
YDD Welfare
Both
YELL0
Both
YELLO
Both
YELLOLTD
Both
you have won
Both
your mobile has been selected
Both
Your number has been selected
Both
Your Number Have Win
Both
Your-Line
Both
ZOOM MULTILINKS
Both
4100
Both
PROMO

Both
reward
Both
Rewarded
Both
18O
Sender
ALERT
Sender
4444555
Sender
44770004
Sender
455445554
Sender
3345566
Sender
3343434
Sender
4456556557
Sender
234
Sender
121
Sender
DIAMOND
Sender
22590409424
Sender
LOADED
Sender

2392290275
Sender
29749
Sender
180
Sender
4017
Both
VIP-CARD
Sender
WWW.LOAD50.COM
Both
Notes:

Both : Defines sender id As well as message content
We regret the inconviniences this might caused you

Thanks for your understanding.
You can use our reliable, cheap bulk sms website.
gtext.com.ng offers you the cheapest bulk sms in Nigeria. Send your bulk SMS on our cheapest bulk SMS portal in Nigeria.
Buy minimum of 10,000 bulk SMS at the rate of N1:40 per UNIT. We offer the best and cheapest bulkSMS in Nigeria.

Check Our **PRICING** Page for more information on **PRICING.**

You can make your payments into the account details below:
Account Number: 0050464651
Account Name: Gileadbalm Enterprises
Bank Name: GTBank

HOW TO MAKE PAYMENTS ONLINE
1.) Click on BUY SMS
2.) Type in the total unit you want to buy in the 'Total' box
3.) Click on 'Vogue Pay' and submit (You will be redirected to the next page)
4.) Specify the preferred mode of payment (Interswitch/Verve/Visa card), then proceed
5.) Enter your email address and click on "make payment"
6.) Specify your card type, fill in the details and click on "pay"
We offer bulk SMS at the cheapest price to all Nigerians on this bulk SMS website. We assure you that you SMS will be sent though the most reliable and fastest bulk SMS gateway which has been serving our numerous bulk SMS clients across Nigeria. We sell the cheapest bulk SMS in Nigeria without compromising the quality of our bulk SMS delivery with our focus been on delivery speed. We don't just offer cheap bulk SMS in Nigeria but we offer fast, reliable and cheapest bulk SMS in Nigeria which all our satisfied bulk SMS users can attest to.

Our system is simple to navigate and we render effective, affordable and reliable bulk SMS services. Try us TODAY and you will definitely not regret it.
We offer you the cheapest bulk sms in Nigeria.

Important Information
Please not the following:
-The Acceptable number formats are as follows 080..., 23480..., +23470...
Please do not use number without the leading zero or without the international country code (eg 8034521...). Message sent to number like this are sent to international numbers starting with 803... and will be charged accordingly.
-To send to multiple numbers, separate multiple numbers with comma "," eg (08032333312,234805849444) or one per line eg
0803422334
0813332232
-To upload or paste numbers from a file please not the following. You may upload numbers but only upload from a notepad file.
Do not upload from MS word, MS excel, MS wordpad or any other document type. If your numbers are contained in any of such file, please copy and paste in notepad first, then upload or copy and paste from notepad. Numbers uploaded directly from documents not notepad are usually not properly formatted.

Our cheapest bulk SMS portal will give you maximum satisfaction and you will definitely continue using our bulk SMS service after your first trial. Any delay experienced sending bulk SMS on our bulk SMS website or other related issues is due to bulk sms delivery issues with GSM service providers. Our bulk SMS gateway is efficient and you are assured of 99.9% bulk SMS delivery rate and speed.

You can send cheapest bulk SMS to all GSM and CDMA networks across Nigeria and networks in all the countries in the world. Smsclone.Com assures you of world class bulk SMS service. Our bulk SMS units do not expire and the bulk sms units remain in your smsclone account. SMSclone's focus is to offer you the best bulk SMS gateway to satisfy your messaging needs. Enjoy the cheapest bulk SMS in Nigeria and the most reliable and fast bulk SMS in Nigeria.

Check Our PRICING Page for more information on PRICING.
Through our cheap bulksms website, you can send message to friends, family, clients etc

PRICING INFORMATION

A page of text message to Etisalat cost 1.4unit, while a page of text message to Glo is 1.6unit, MTN and Airtel cost 1.5 units of SMS. Please see the GSM networks and unit charged below for better clarification.

GSM Networks	Unit(s) of sms for a page of text message
Airtel	1.4 Units
Etisalat	1.4Unit
Glo	1.4Units
MTN	1.4 Units

CHAPTER FOUR

BULK SMS RESELLER PACKAGE

CHAPTER FOUR

BULK SMS RESELLER PACKAGE

We can develop a bulk SMS website that will bear your own domain name and you will have 100% control over it. You will be able to customize it, edit it and determine the prices to sell to your own users as well as your own resellers.

For more information or order placement, please us via email or telephone. You can make payment for SMS credit into our bank account.
Sales Letter!

Importance of Sales letter cannot be over emphasize. It must however be writer by a professional and must be real and authentic. If you write a sales letter that is not true and consistent with your experience, you will end up losing out and will not be able to give tremendous value which is the only reason that will cause repeat patronage.

Below is sample of one of the sales letters I used last year;

" EXTENDED OFFER: NIGERIA'S TOP MOBILE MARKETER TO GIVE-AWAY HIS COMPLETE MOBILE MARKETING SYSTEM THAT GENERATED OVER N100,000,000 INCOME IN 2015

This OFFER has further been extended to Friday January 29th 2016.

Big rush in Nigerian mobile Marketing industry as the nation's Top Mobile Marketer gives up ALL His Trade Secrets and Resources In The Biggest Mobile Marketing Retirement Fire Sale Nigeria Has Ever seen!

It's no News that I Have personally RETIRED from Marketing Bulk SMS & Databases Since October 1st to focus on 8 New Online Sites our company is building. I have made Hundreds of Millions from bulk SMS business. I have service Presidential candidates, Governors, Senators and Top Multinational such as Chivita, PZ, Carex, just to name a few. Need to start business I can take to Silicon Valley get investors and make it global and need focus to do this but will gladly release all my secret in the Bulk SMS business and Mobile Marketing Business to you. I Did a training on October 1st where I share my secret of playing at such a big level.

BELOW ARE SOME OF THE THINGS I THOUGHT PERTICIPANTS FOR FREE;

1) How to make 7 Figures from Bulk SMS without physically seeing the client
2) How to Build an Online brand that cutoff physical meeting before your client make payment
3) How to Automate your bulk SMS business
4) The New Way to sell GSM database and not be on the wrong side of NCC
5) How to Purchase quality database from genuine sources
6) Why database is the main backbone of BulkSMS business
7) How to make transaction with multinationals without writing a single proposal
8) How to use email marketing to market your bulkSMS & database biz
9) How to use FREE automated social media campaigns to market your BulkSMS business and get direct convention.
10) How to become a super dealer and have resellers under you
11) How to prevent 10 Million Naira loss from bulk SMS business like I did
12) Add you to my secret Facebook group where all your bulk SMS questions can be answered
13) How not to loose million from bulk SMS unit thieves who stole 10Million units from my friends website

The whole of the above am giving Free at a Seminar By Filling below or visit our office, 4th Floor 52 Ijaye Road beside UBA Bank, Ogba, Lagos. Number to Call is 2348188220077
HOWEVER,

1) The Major secret that will help you make the kind of money I made and to consult for the kind of multinational I consulted for was simple because I also paid millions to acquire the over 10 GSM& Email Database we have and resold it at higher amount and made crazy profits from it. So if I teach you and teach you and don't give you the database, it will be hard to attain my level.

2) Due to Security reason, you are advice not to sell database any how to anybody but I will advise the use of a new spc4 joomla website that will hide from the customer the full number so that clients can extract the number but can just send message to them. Since Nigeria doesn't have anti-spam law yet. Hence you need a higher version of Joomla and SPC component site.

3) Mentors are guide, with a good mentor there is no limits to your greatness. We have a very arrogant generation and that is why many of us are poor. We all think we can just get things without looking for a good guide to show us. If I have made millions and can't produce millionaires, then am a failure but thank God that I have already produce millionaires and just want to produce many more?

Hence below offer is for a FEE. My Mentor Bishop Oyedepo has always said; "Nothing of value is free". I absolutely agree. No one value what is given to him or her free. No one! Jim Rohn is one of my favorite Mentors and teacher. I recently listen to one of his protégés by the name ANTHONY Robbins share how he met Jim Rohn, according to him, he was age 17 and was earning 40dollars a week when he met Jim Rohn and told him to teach him how to be successful and Jim agreed but said to come for my seminar where I am going to teach you, it will cost you 35dollars. Anthony Robbins was shock but Jim insisted he paid.

That little investment has made ANTHONY Robbins one of the biggest speaker and consultant in the world today. Anything of value will cost you something.

Below is the package I will charge you for;

OUR DATABASE

Category A:

1) Get the Most Complete GSM Database classified by (Name, Sex, LGAs, State) for the whole Country.100% Satisfaction guaranteed.

Nationwide database valued at 450,000naira.

2) Lagos State Street by Street database valued at 100,000.

3) 10,000 business Directory (email, Phone number &addresses)- 45,000naira.

4) 500,000 email addresses of online users - 120,000naira

5) Get 50,000 email addresses and 50,000 phone numbers of senior and cooperate executives 55,000naira

6) 880,000 working class email-55,000naira

7) 5000 Surulere Business Owner Database 45,000naira

8) 7000 Ikeja Business Owner Database 45,000 naira

9) 10,000 Island Business Owner Database- 75,000naira.

10) 600,000 email address of E-Commerce Shoppers worth 180,000naira.

11) Spc4 component Bulk SMS website.

12) Online Payment Integration and Social Media Integration

13) My bestselling book on Amazon.com "The Mobile Millionaire" inside it is my popular snook; How I made 3.5 Million in two weeks.

14) This one is the most important and that is 6month online and offline Mentorship worth 1million naira.

If you are Outside Lagos, or Not In Nigeria, This is a business you can do from any part of the world. We have 5 Videos that train you on A TO Z Of this business. The Video on How to make 7Figure in bulk SMS Business video and 4 other video that explain how to design a reseller website and how to market your business and manage the business.

All these my retirement gift shouldn't be less than 5,000,000 but I will not sell it for half of that. No! Will not even give them to you for 500,000naira.

But Pick any 1 for 15,000naira but the price for picking one item increase by 5000naira daily. or 3 of these list for 50,000naira and any 7 for 100,000naira and the whole 14th Products for 200,000naira only.

NOTE: 3package, 7package and whole package ATTRACTS WEEKLY INCREASE OF 25,000Naira.You are advice to get it now. We have limited number of people we take for this extension due to the demand website design. Even though our IT Team is more than 10 but the demand is high. Also because I can't mentor large number of people for 6months I have other people who I mentor who paid 1million each.

This offer ended 20th November. However we are not going to extend further. Since October 1 I retired from bulk sms to focus on my new business. The Best am doing now is to mentor people and train them to be successful in that business. I Am NO longer personally involved in anything bulk SMS. I have trained enough staff and resellers who will handle our bulk SMS business successfully even beyond the level I did.

To Order,
1) Fill the form below
2) Make Payment to GTBank, Gileadbalm, 0050464651.
3) Call Mariam on 2348188220077. and fill the payment form so that your packages will be sent to your email address.
The Link to the payment form after your have paid is: http://bit.ly/paymentmade note that this link is only to be filled after payment is made.
For physical meeting you can visit our office and talk to any of our staff.
 Office: 52 Ijaye road,Beside UBA Bank,Ogba.Lagos.
Stephen Akintayo
CEO Gileadbalm Group.
Linkedin/ Stephen Akintayo
Facebook: stephenakintayonigeria
Instagram stephenakintayo
Twitter: @stephenakintyo
www.stephenakintayo.com
www.gileadsms.com "

My advice to you is to avoid being unnecessarily hasty. Bulk SMS business is a residual business. So you need time to build its residual nature. It might take years to build. If you can't wait its better you quit now. Take it one step at a time. You will get there if only you know how. Please forget the time it might take you, the important thing is that you will get there.

In addition, my advice to you is to be focused on building your customer base. Set your target and work towards it gradually. Imagine if you can win one new customer per day. In one month you would have gotten 30 new customers, if you do that in 90 days that's about 90 customers. However, it's not really as easy as that. However, use it as your guide.

Important questions to ask before starting your bulk sms business?

Why GSM Database?

Do you have a product or service you want to market to Nigerians or you have an announcement to make, whatever your needs we have answers to the most effective means of reaching millions of Nigerians. With us you can reach most of the Nigerian population instantly.

Very Important:
Hundreds of people spend between N20, 000 – N3,000, 000 advertising in local newspapers, Radio and Television stations daily thinking that their products will be seen by hundreds of thousands of people.

If you are one of them, then here is the simple truth. Except there is breaking news of national interest, hardly will you find 100,000 people reading a newspaper a day out of the 150 million Nigerians. Isn't that shocking?

And many of those who even decide to read, due to insufficient time in their hands, skip pages of the newspaper to the segment they are interested in. E.g. Entertainment and show biz, sports etc. Out of 100 people who advertise on such mediums, only about 2 percent come back to say they have made 10 – 20 sales.

Why BULK SMS?
Current statistics:
9% of Nigerians watch TV daily
19% of Nigerians listen to radio daily
4% of Nigerians read newspapers daily
29% use the internet daily
50% of Nigerians use GSM phone daily

phone numbers from every state throughout Nigeria sorted according to the 36 states in Nigeria including FCT Abuja and up to 90% active phone lines with periodical updates. So you can rest assured whatever message you are passing through will be received all over the country instantly.

We have over 21 million fresh Nigerian phone numbers in our database.

This is a limited time promo. (That is less than 1 kobo per Number)

We will return to the normal price at anytime!

Limited Copies available!

Conditions of Sale

The Database is NEVER to be given out for Free (we would cancel your access if this is discovered)

A minimum price that it can be re-sold would be sent to you with the Database

This database must not be used for SPAM, Criminal or Illegal Purposes (we ban users daily on our sites for spam)

Here are a few questions to ask your would-be SMS provider:

1. Will I have complete control over my website?

This is because you need to enjoy complete control of your website. A good reseller platform will provide you with access to your c-panel and database as well.

2. Will my site look exactly like yours?
If yes, run! You should always strive to have a unique looking website. Nobody likes a photocopy of the original.

3. Will my site have a mobile version or mobile theme?
This is important because as at today, 62% of people in Nigeria access the internet from their mobile devices. And if your website is not accessible on their phones, they will go to another person to do business. Aside from your normal site, there has to be a mobile optimized site for your mobile phone users.

4. Is it simple as a-b-c, If I choose to, can I edit my website and add other services?
E.g. if you decide to sell other products, expand it into a dating site or add a forum.

Here are a few questions to ask yourself.
1. Can I make it easy for my clients to make payment for their purchase on my website?
If your buyers have to physically go to the bank or come to you before they pay for your SMS services, you will be out of business in no time. Hence, your portal should support a method for online payment.

2. How easy is it for me to make payment to my provider?

As a reseller of SMS credits, you will want to enjoy an easier way to buy SMS from your provider. You want your provider to provide you an easier way of making payment so that you can deliver to your client with comfort and without stress. If you have to go to the bank every time before you buy SMS, you will be out of business very shortly.

3. Can you conclude that the delivery rate of their SMS is satisfactory?
SMS messages are delivered on priority. The good ones guarantee immediate delivery of the messages. That is what you need to make your clients happy
Why are all these questions important to your new business?

The truth is you are about to invest your own money on a business system and you must carefully look for a provider that can provide all the basic requirements and even surpass your expectations. Also, you do not want to experience break down in your service delivery and want to satisfy your clients. These questions should serve as a preparatory guide that would save you a lot in terms of money and clients in the long run. There's no substitute for preparation.

When you successfully answer these questions, you will have covered the two basic business systems needed to run your SMS business, namely (a) SMS gateway and reseller website and (b) Online payment platform.

If you already have a bulk SMS website and you feel the need for a better portal or SMS Provider, I suggest you take a courageous step now than regret later. You should demand for these features from your provider, or better still, talk to another provider and begin preparation for moving your portal.

Sample of Frequently Ask Questions About GSM Database?

Read the Frequently Asked Questions & Their Answer Below...

Q: In what file format is the GSM numbers?

The numbers are stored in Microsoft Excel which every computer has and can open.

Q: How genuine are the numbers?

You can do your own due diligence by calling some of the numbers to verify them when you get them.

Q: How many states are covered?

The whole 36 states of the federation including the Federal Capital Territory, Abuja

Q: Is it really up to 80 million numbers?

Yes. Microsoft Excel automatically gives you the total number, so you can verify it on your own when you get access to it.

Q: What are the networks covered?

The four major networks: MTN, GLO, Airtel, and Etisalat.   Q: How much is it?   Our normal price is N100,000 but in the spirit of the season, we're giving you a 25% cash back. That means you pay just N75,000 while this special offer lasts. Offer expires by midnight December 31st, 2012.

Q: Why is contact all SMS selling this millions of numbers for just N75,000 when others are selling it for as high as N350,000; is yours for real?

Yes, ours is for real; it's the same numbers we are all selling. We are selling this low because we are doing our BTD Promo. Once the promo is over, the price will go back to normal: N100,000.

Q: Why sell to only 10 people?

Because we don't want to drive down the worth of the GSM numbers by putting it into the hands of too many people. We also don't want the numbers to get into the wrong hands.

Q: If I pay for the numbers, how do I get them?

We will email you a link to download it. If you want it on DVD, it will cost you a little extra: N5,000 (and that includes shipping fee, meaning that we will deliver it to your doorstep).

Q: How do I make payment?

Pay into any of our bank below, you can use branch locator to locate closest branch to you.

REFERENCES

REFERENCES

1. 8Ernst & Young, 2012. "Tracking global trends: how six key developments are shaping the business world"
2. www.internet.org
3. GSMA, THE MOBILE ECONOMY 2013
4. http://www.businessinsider.com/complete-visual-history-of-cell-phones-2011-5?op=1#ixzz2xGPgO6R4
5. http://mashable.com/2013/08/05/most-used-smartphone-apps/
6. http://en.wikipedia.org/wiki/Mobile_marketing
7. www.visiongain.com
8. http://www.jwtintelligence.com/wp-content/uploads/2013/04/F_JWT_13-Mobile-Trends-for-2013-and-Beyond_04.02.13.pdf

Made in the USA
Middletown, DE
19 September 2024